ISSN 1020-12[redacted]3

Water Reports 21

WATER QUALITY MANAGEMENT AND CONTROL OF WATER POLLUTION

Proceedings of a Regional Workshop
Bangkok, Thailand
26-30 October 1999

FOOD AND AGRICULTURE ORGANIZATION OF THE UNITED NATIONS
Rome, 2000

ISBN 92-5-104503-8

Contents

Summary report, conclusions and recommendations

SUMMARY REPORT

The Expert Group emphasized that water is a key resource for all economic sectors which increasingly must compete for this scarce commodity. Therefore, every effort must be taken to mobilize personnel, facilities, financial resources and the application of new practices and technologies to facilitate the efficient and effective use of all types of water (freshwater, brackish, saline, wastewater) in agriculture. An integrated and comprehensive approach is essential for water allocation and management both for water quantity and quality.

Use of wastewater in agriculture has substantial benefits in the Region. However, substantial change in practices are required to safeguard public health and the health of agricultural workers, to ensure efficient water use, and to control salinity and off-site water pollution.

Protection of water quality is a multi-jurisdictional issue, however national water programmes suffer from lack of policy focus for water quality. Modernization of water quality policies and programmes are required to prevent institutional and policy failure, and to ensure effective delivery of data programmes that have benefit to agriculture and other sectors of the economy.

CONCLUSIONS

For beneficial use of **wastewater** in agriculture, the following were identified as the principal issues:

- **The need to exploit the potential of treated wastewater and excreta as valuable resources:** Current practices vary within in the Region, however wastewater use rarely complies with the required health and environmental safeguards. Environmentally sound use of wastewater and excreta could lead to enhanced food and feed production, conservation and enhancement of water resources, reduction in use of agro-chemicals, and in improved health and environmental benefits.
- **Salinity:** This is a serious problem which affects yield and may cause deterioration of surface and groundwater quality with adverse and potentially irreversible effects on soil and the environment.
- **Low water use efficiency:** This is one of the main factors that reduces water productivity and increases the potential for environmental pollution in agricultural areas.

For control of **water pollution** from agriculture, and for use of polluted water in agriculture, the following were identified as the principal issues:

- **Policy failure**: Water quality management is first and foremost a problem of policy which in many countries fails to recognize the national significance of water quality, the cost of degraded water to the national economy and to the agricultural sector, and the loss of environmental benefits from water pollution. A modern policy response to water quality management is essential but is lacking in most national water resource policies.

- **Institutional and legal reform:** These are key elements in water quality management and include a better definition of roles and responsibilities of institutions, and a cooperative framework for water quality management. It also includes a legislative response to efficient and enforceable regulations to ensure data quality, the regulation of effluents, and the definition of achievable and realistic water quality standards and objectives.
- **Data programmes:** are a key element in development of modern water management policies, in planning and management, and in decision-making on water quality remediation investments. However, data programmes in most countries are inefficient and often provide little information of value to the policy maker or to the regulator. For agriculture, water quality data programmes rarely take into account the specific needs of agricultural agencies for information required to develop and evaluate farm management practices that will improve off-site impacts on water quality.
- **Management practices:** These are a key element to achieving improved off-site and groundwater quality from agricultural activities. Greater accountability is required of the private sector in its role in agro-industries and agricultural production.
- **Capacity building:** National governments often adopt an uncritical approach to donor programmes that often focus more on what the donor perceives to be needed rather than on what the country may actually need. Consequently, there are many examples of unsustainable technologies that are introduced into national programmes that fail when the donor withdraws. There must be a more focused effort to transfer technologies (hard and soft) that are sustainable in the environment of the receiving country. Concern was also expressed over the use of foreign experts who often leave little improvement in local capacity, and in the use of tied aid that focuses on the support of the donor's private sector rather than necessarily on what is actually needed in the receiving country. At national levels, much greater use could be made of local expertise however this is often frustrated by competition amongst national agencies for donor support. Use of modern information technology tools, such a decision support systems that bring knowledge and expertise into the hands of decision-makers, can be very efficient and effective.

RECOMMENDATIONS

The following recommendations were made for implementation by local, national, regional and international organizations:

Wastewater use in agriculture

Public policy

- It is recommended that the guidelines developed by WHO/FAO/UNEP for use of wastewater in agriculture be promoted as a basis for preparing national guidelines, regulations, and codes of practice. Governments should consider providing incentives to industry to treat effluents to the minimum level that is recommended for use by the agricultural sector.

Health aspects

- Protecting public health and the environment are the main concerns associated with wastewater use. The health and environmental risks should be within acceptable levels. A minimum treatment is required to achieve acceptable purified wastewater that is safe for designated uses. Raw wastewater is not recommended for any irrigation purpose.

- There are apparently no controls in the countries of the Region over the crops grown which are irrigated with wastewater. It is recommended that vegetable crops, normally eaten raw, should not be irrigated with inadequately treated wastewater. Wastewater may be primarily used in agro-forestry, orchards, cereal and industrial crops. To prevent workers from wastewater exposure, they should use footwear and gloves and utilize appropriate methods of irrigation and sludge application. Farmers need to be more awareness of these facts.

Environmental aspects

- Wastewater irrigation can enrich the soil with organic matter and nutrients (N, P, K) and increase its water holding capacity and it may increase crop production. However, urban and industrial wastewater may also contain toxic chemicals like heavy metals. Long term uncontrolled use of wastewater may lead to a build-up of soil salinity, accumulation of toxic chemicals and reduction of soil permeability, and pollution of surface and groundwater. General guidelines on irrigation water quality (FAO) should be applied to avoid immediate, short and long term detrimental effects on the environment. Monitoring of ground and surface water resources close to the wastewater area should be carried out regularly to provide an early warning of pollution status and risks.
- Promoting use of wastewater in agriculture as an alternative to discharge to surface waters will decrease potential for eutrophication of surface waters.
- In order to control pollution of water bodies due to disposal of saline drainage, efforts should be made to minimize drainage surplus by resorting to methods for increasing water-use efficiency.
- Treatment procedures for waste purification which are cost effective and environmentally friendly should be promoted.

Legal and institutional aspects

- Develop, amend, and/or adopt legislation which will enable the appropriate use of treated wastewater and excreta. Compliance with the legislation needs to be enforced. National action plans should be prepared to include, among others, institutional framework, inter/intra sectoral co-ordination, human resource development and technology options.

Socio-cultural aspects and human resources development

- The socio-cultural aspects of wastewater reuse should be examined before planning local wastewater systems; Women should be actively involved in all phases.
- Public awareness at local, national and regional levels should be promoted through increased dissemination of information through public media as well as at workshops, seminars and exchange of visits.

Research and development, technology transfer

- Gaps in knowledge and information should be identified and research proposals prepared for submission to national and international agencies for support. Adequate funding is essential.

Health, agricultural, environmental, and ecological implications, and various issues related to operation, maintenance, and management need to be studied in-depth in order to develop regional strategies and country-specific norms.

- A regional network for this sector should be established to promote information and experience regarding relevant research and technologies, and to promote exchange and co-operation among the countries of the region and with UN organizations.

Control of water pollution

Public policy

- Modern water policy formulation must explicitly include water quality concerns. Policy reform for water quality management should include clear objectives and an action plan for implementation.

Institutional issues

- Institutional and legal reform is required to bring institutional efficiency and modern legal and enforceable regulations into the management of water quality. Important elements include new and enforceable approaches to effluent control, the legalization of national data standards, and the evaluation of new or alternative institutional arrangements that make better use of public-private sector partnerships.

Programme reform

- Water quality programmes, including monitoring and data programmes, need to be modernized both to take into account new technical developments in efficient water quality monitoring and assessment, and also to increase efficiency and effectiveness in these programmes so that they respond to real data needs by data users and by decision-makers.

Management practices

- Management practices can be greatly improved to increase efficiencies in water use, to reduce use of agrochemicals, and to reduce of off-site impacts on surface and groundwater quality.

Capacity building

- Capacity building is essential, but needs to be more carefully considered both by donors and recipients to ensure that there are real gains in capacity and the new tools and techniques are sustainable within the socio-economic fabric of the receiving country.

These recommendations are amplified in much greater detail in the general report of the Expert Meeting.

Follow-up Recommended

The following are recommendations for follow-up to United Nations specialized agencies, other multi-lateral agencies, donors, and to national governments, in the field of wastewater reuse and for control of water pollution:

1. Promote and assist in the modernization of policies, regulations, laws and programmes in water quality management and wastewater reuse, and to encourage governments' commitment to this objective. This should include transparency of government policies and programmes and accessibility to data.
2. Capacity building needs to focus more on developing local expertise and on sustainable practices so that foreign "experts" increasingly become facilitators rather than "doers". Conduct training courses, seminars and workshops at local, national and regional levels in different countries. Assessment of training needs and existing training facilities should be made. Existing curricula in the field of agriculture, aquaculture and forestry, health, engineering and water resources management should be strengthened and modified to include issues related to control of water pollution and to wastewater use. For the in-service workers and managers, continuing education and non-formal short term training modular courses should be instituted.
3. United Nations organizations should facilitate transfer of appropriate and cost effective technologies and to help with their adaptation under local conditions. This should include programmes that allow countries to share experiences, lessons learned, and which promote technical cooperation amongst developing countries,.
4. United Nations agencies need to more carefully coordinate and rationalize their collection of data on water quantity and quality, water use and reuse, and related issues in the Region and globally, and to ensure accessibility to these data sets.
5. Carry out a selection of case studies in representative countries of the region in order to elaborate the benefits, dis-benefits and best management practices that can be applied to the region in wastewater reuse and management, and for control of water pollution.

Annex 1
Opening address

Distinguished Participants, Ladies and Gentlemen,

On behalf of the Director-General of FAO, Dr. Jacques Diouf and on my own behalf, I have the honour to welcome you to the Regional Workshop on "Water Quality Management and Control of Water Pollution in Asia and The Pacific".

I am grateful that you accepted our invitation to attend this important meeting. The presence here of a range of international and national experts, engineers as well as senior administrators concerned with the various aspects of water quality management and water pollution prevention, indicates the keen interest being taken by the various countries of the region and the international organizations in this important subject.

I am particularly pleased to welcome you here on the premises of our Regional Office in Bangkok. I hope you have a pleasant stay and a fruitful exchange of ideas.

We are now approaching the second millennium which brings with it the unparalleled concern over the scarcity of freshwater resources to meet the future demand of our growing population. Overall, the region is relatively well endowed with water resources. However, if we look at the amount of water available per inhabitant, the situation is less promising. In terms of water resources per person, the figure is just slightly above half of the world's average. Countries of the Indian subcontinent, Eastern Asia and the Far East show the lowest figures while Southeast Asia has still more water resources per person than the world average. The figure of 2000 m3/inhabitant/year is usually used as an indicator of water scarcity: India and China are approaching this limit, while the Republic of Korea is already below it, at 1538 m3/inhabitant/year. The situation is even more alarming when we look at most of the small islands in the Pacific or the Maldives where, in the absence of any surface water the growing population is solely dependant on limited groundwater resources which due to over pumping are already subject to seawater intrusion with subsequent deterioration of its quality.

While recent droughts have significantly increased the awareness of many countries in the region that scarcity of fresh water resources will already become a serious issue in the next decade when the world population is expected to exceed 7 billion people, the alarming deterioation of the water quality, however, is still not attracting sufficient attention and appropriate counter-measures

At present water quality is being affected by harmful effects caused by many sectors of human activity. Agricultural, industrial, mining and urban by-products and residues are the principal pollutants of inland waters and oceans. Less than 2% of the cities of the world have treatment plants, and in rural areas most agro-industries, dairies, feedlots and small communities simply dump their wastes into streams and other water bodies polluting them with nutrients, bacteria and parasites. The World Health Organization (WHO) carried out a survey in 1989 on 110 rivers throughout the world and found that in 49 of them, the faecal coliform levels exceeded the recommended WHO level. Also in Asia and the Pacific, 15 out of 32 rivers tested exceeded the

Prem Nath
Assistant Director-General/Regional Representative for Asia and the Pacific

faecal coliform guideline significantly. Now, 10 years later the results would be even worst if the survey was repeated.

As a consequence of this pollution, researchers have estimated that 80% of all diseases and 33% of deaths in developing countries are linked to poor water quality. The outbreak of cholera in Peru in 1991-92, in addition to the loss of countless human lives, was estimated to have cost US$460 million in lost exports and tourism. Furthermore, over-exploitation of groundwater resources has resulted in the extraction of water from deeper aquifers with sometimes high arsenic content which has a serious toxic effect on human beings. Water with high levels of arsenic are the only source of drinking water in a number of cities and rural communities of the world. Well documented examples of arsenic poisoning are reported from Chile, China, Mexico, Peru and Bangladesh.

As already mentioned, there are several sources that contribute to water pollution and these may be broadly grouped into agricultural sources and non-agricultural sources. In this connection, it should be noted that agriculture, in addition to being the largest consumer of water, is also listed, with increasing frequency, as a major contributor to water pollution. Irrigation return flows to surface and groundwater resources contain heavy loads of salts, nutrients and agro-chemicals, thus contributing essentially to the deterioration of water quality.

There is still ample room for politicians and decision makers all over the world to pay greater attention to the increasing imbalance between our steadily growing population and the limited but not renewable natural resources at our disposal in particular soil and water. This imbalance, which has already caused past difficulties and problems in the conservation of specific ecosystems, has nowadays reached unexpected dimensions in many places of the world. There are clear signs of exhaustion of natural resources. In relation to water, there are serious pollution problems not only with inland waters - most closely connected with the population - but also with the oceans, which have always been considered unlimited and inexhaustible because of their immensity. Symptoms of deterioration are not only evident in Exclusive Economic Zones or in the continental shelves, but also on the high seas, whose resources are currently being affected.

FAO, together with many other institutions, has been working for many years to raise awareness on the issue of deteriorating water quality as well as on the identification of causes and the proposal of solutions to prevent or at least to minimise water pollution. This has been achieved through numerous technical publications, the conduct of seminars and expert consultations with the objectives of promoting an exchange of experiences on water pollution by agriculture, agro-industry and mining and in particular on the safe use of wastewater and highly saline waters for irrigation, which in view of the increasing water scarcity will gain more and more importance in the future.

A first regional workshop on the subject and organized by FAO was held in Chile in 1998, gathering participants from the Latin-America and Caribbean Region. This second workshop focuses on Asia and the Pacific. The presence of specialists from the various national and international agencies and institutions in this Workshop will no doubt provide an important opportunity to exchange and consolidate experiences and to indicate ways for further follow-up and regional collaboration.

I am confident that with your active participation and technical contributions, solutions to the many water quality management problems in the region will be proposed in order to protect and to sustain our scarce water resources for the generations to come. FAO is happy to be a part of this, and I am anxious to learn of your reactions and recommendations at the end of your three days of deliberations.

I wish you a very successful meeting and a pleasant stay in Bangkok.

Thank you.

Annex 2
Provisional agenda

Date	Time	Activity
Tuesday 26/10/99	08:00	Registration
	08:45	Opening address by Dr. Prem Nath, Assistant Director-General /Regional Representative for Asia and the Pacific. Welcome and general information on the workshop by Dr. Klaus S egert, FAO, Secretary of the meeting
	09:15	Fernando Chanduví, FAO Technical Officer, Water Resources Development and Management Service (AGLW): Results of the Survey on Water Quality and Pollution Control in the Region.
	09:30	Coffee break
		Technical Session I: Water Pollution by Agriculture, Agro-Industries and Mining **Chair: Dr. Ioannis Papadopoulus, Cyprus**
	10:00	**Keynote speech** "Water Quality Management in Asia and the Pacific" by Mr. Edwin Ongley, Canada.
	10:30	**China**, "Quality of Irrigation Water in China" by Mr. Weng Jianhua.
	11:00	**Laos**, "Water Quality Management in LAO PDR" by Ms. Keobang A Keola.
	11:30	**Vietnam**, "Water Quality and Agriculture Production in the Mekong Delta" by Messrs. To Van Truong and Pham Gia Hien.
	12:00	Plenary Discussion
	12:30	LUNCH
		Technical Session I continued….. **Chair:** Mr.**Fernando Chanduvi, FAO**
	14:00	**Malaysia**, "Water Pollution by Agriculture, Agro-industry and Mining in Malaysia" by Mr. Tan Choo Yong.
	14:30	**Pakistan**, "Agriculture's influence on water quality" by Mr. Bashir Ahmad.
	15:00	**Japan**, "The impact of agriculture on water quality and its improvement methodology in Japan" by Mr. Yoshito Yuyama
	15:30	Coffee Break.
	16:00	A case matching Decision-Support System to predict agricultural impacts on water quality by Mr. Edwin Ongley.
	16:30	Plenary Discussion
	19:00	Welcome Dinner at Siam City Hotel hosted by Dr. Prem Nath

Wednesday 27/10/99		**Technical Session II: Use of Wastewater and highly saline Water for Irrigation** **Chair: Mr. Edwin Ongley, Canada**
	08:30	**Keynote speech**, "Use of Wastewater for irrigation with reference to Asia and the Pacific" by Mr. Ioannis Papadopoulos
	09:00	**Bangladesh**, "Crop Production in the Southern Saline Belt of Bangladesh" by Mr. Md. Abu Bakar.
	09:30	**India**, "Agriculture and Water Quality in India towards Sustainable Management" by Mr. R.P.S. Malik.
	10:00	Plenary Discussion
	10:30	Coffee break.
	11:00	**Indonesia**, "The Biochemical Effectiveness of Herb Bed Treatment System with Bulrush, Water Morning Glory and Garden Cormous Herb in Waste water Improvement" by Messrs. Soedodo Hardjoamidjoyo and M.S. Saeni.
	11:30	**Korea**, "Status of Agricultural Water Quality in Korea" by Mr. Lee Jong-Sik.
	12:00	**Thailand,** Agriculture's Influence on Water Quality in Lower Northeastern Thailand" by Ms. Vipada Apinan
	12:30	Lunch
		Technical Session II continued..... **Chair: Mr. Bashir Ahmad**
	14:00	**Keynote speech**, "Organic Sewage Treatments" by Mr. Maurizio Giannotti
	14:30	**Philippines**, "Modelling Non-point Source of surface and ground water systems in selected agricultural watersheds in the Philippines" by Messrs. R.S. Clemente and E. Wilson.
	15:00	**Thailand,** "Utilisation of Spent Distillery Wash Liquor in Paddy Fields" by Mr. Suchint Phanapavudhikul.
	15:30	Coffee break
	16:00	"Phytodepuration treatment of wastewater from agro-industries to avoid water pollution" by Maurizio Giannotti.
	16:30	Plenary Discussion
Thursday 28/10/99	08:30	Work in two groups drafting (i) conclusions and (ii) recommendations
	10:00	Coffee break.
	10:30	Group Work continued
	12:30	Lunch
	14:00	Video session
	16:00	Break
		Closing Session: Chair: Dr. Klaus Siegert, FAO
	16:30	Presentation of Conclusions and Recommendations Adoption of workshop report Close of Workshop
Friday 29/10/99	07:00	**Field trip**

Annex 3
List of participants

BANGLADESH	Mr. M. Abu Bakar Deputy Director Field Service Wing Department of Agricultural Extension Khamarbari, Farmgate Dhaka 1215
CHINA	Dr. Weng Jianhua Senior Engineer Water Quality Research Centre China Institute of Water Resources and Hydropower Research 20 Chegongzhuang West Road P.O. Box 366 Beijing 100044 Tel: 86-10-6841-5522 Ext. 3203 Fax: 86-10-6842-5954 E-mail: wengjh@public.east.cn.net
INDIA	Dr. R.P.S. Malik Agricultural Economics Research Center University of Delhi New Delhi 110007 Tel: 91-11-7257648 Fax: 91-11-6103987 E-mail: malik@aercdu.ren.nic.in
INDONESIA	Prof. Soedodo Hardjoamidijoyo Bogor Agricultural University P.O. Box 220 Bogor 16002 Tel/Fax: 0251 627225 E-mail: soedodo@wasantara.net.id
JAPAN	Dr. Yoshito Yayuma Expert of Water Management The Modernization of Water Management System Project JICA, Irrigation Engineer Center Royal Irrigation Department 811 Samsen Road Bangkok 10300 Tel: 2431098, 2431318 Fax: 2431317

REPUBLIC OF KOREA

Dr. Lee Jong-Sik
Researcher
National Institute of Agricultural Science and Technology
249 Soedun-dong
Kweonseon-ku, Suwon
Tel: 82-331-2900214
Fax: 82-331-2900277
E-mail: jongslee@niast.go.kr

LAO PDR

Mrs. Keobang A. Keola
Deputy Chief of Cabinet
Prime Minister's Office
Science Technology and Environmental Agency (STEA)
P.O. Box 2279
Vientiane
Tel: 856-21-213472
Fax: 852-21-213470

MALAYSIA

Mr. Tan Choo Yong
Project Engineer
Department of Irrigation and Drainage
Pejabat Jurutera Projek
IADP Seberang
Perak/Tumboh Blok
Jabatan Pengariran dan Slairan
Sungai Dedap
3600 Kampong Gajah
Perak
Tel: 05-6551362
Fax: 05-6552208
E-mail: jpssptb@tm.net.my

VIET NAM

Mr. Pham Gia Hien
Water Quality Laboratory
Sub-Institute of Water Resources Planning and Management
Ho Chi Minh City
Fax: 84-4-8351721
E-mail: giahien@hem.fpt.vn

RESOURCE PERSONS

CANADA

Mr. Edwin D. Ongley
Senior Advisor
United Nation's GEMS/Water Programme
3486 Burlington
Ontario L7M 1Z6

CYPRUS

Mr. Ioannis Papadopoulos
Agricultural Research Institute
Nicosia
Email: papado@arinet.ari.gov.cy

ITALY

Mr. Maurizio Giannotti
Waste Water Treatment Consultant
Viterbo
Email: sifvt@tin.it

OBSERVERS

WHO

Dr. Kanchanasak Phonboon
WHO National Consultant on Health Promotion
Environment and Health Programme
Office of Environmental Policy and Planning (OEPP)
Soi Pibulwatana 7, Rama VI Road
Bangkok 10400

Ms. Raweewan Ratanavaraha
Environmental Officer
Natural Resources and Environmental Management Division
Office of Environmental Policy and Planning (OEPP)
Soi Pibulwatana 7, Rama VI Road
Bangkok 10400

Mr. Pirun Saiyasitpanich
Environmental Officer
Environmental Impact Assessment Evaluation Division
Office of Environmental Policy and Planning (OEPP)
Soi Pibulwatana 7, Rama VI Road
Bangkok 10400

GTZ

Dr. Gustav M. Maurer
Principal Adviser/Team Leader
Agricultural Toxic Substance Division Building (4th Fl.)
Department of Agriculture (DOA)
Chatuchak, Bangkok

AIT

Dr. Preeda Parkpian
Coordinator
Environmental Engineering Programme
Asian Institute of Technology
P.O. Box 4, Klngluang
Pathumthani 12120

UNICEF

Dr. T.V. Luong
Consultant
19 Phra Atit Road
Bangkok 10200

FAO

Mr. Mathias Halwart
Technical Officer
Inland Water Resources and Aquaculture Service (FIRI)
Fishery Resources Division
FAO, Rome
Iraly

FAO SECRETARIAT

Dr. Klaus Siegert
Secretary of the Workshop
Regional Water Resources Development Officer
RAP/FAO
Bangkok 10200

Tel: 2817844 ext. 298
Fax: 2800445
E-mail: klaus.siegert@fao.org

Mr. F. Chanduvi
Technical Officer
Water Quality and Environmental Impact Group
Water Resources, Development and Management Service
AGLW
FAO, Rome
Italy

FAO TECHNICAL PAPERS

WATER REPORTS

1	Prevention of water pollution by agriculture and related activities, 1993 (E S)
2	Irrigation water delivery models, 1994 (E)
3	Water harvesting for improved agricultural production, 1994 (E)
4	Use of remote sensing techniques in irrigation and drainage, 1995 (E)
5	Irrigation management transfer, 1995 (E)
6	Methodology for water policy review and reform, 1995 (E)
7	Irrigation in Africa in figures/L'irrigation en Afrique en chiffres, 1995 (E/F)
8	Irrigation scheduling: from theory to practice, 1996 (E)
9	Irrigation in the Near East Region in figures, 1997 (E)
10	Quality control of wastewater for irrigated crop production, 1997 (E)
11	Seawater intrusion in coastal aquifers – Guidelines for study, monitoring and control, 1997 (E)
12	Modernization of irrigation schemes: past experiences and future options, 1997 (E)
13	Management of agricultural drainage water quality, 1997 (E)
14	Irrigation technology transfer in support of food security, 1997 (E)
15	Irrigation in the countries of the former Soviet Union in figures, 1997 (E)
16	Télédétection et ressources en eau/Remote sensing and water resources, 1997 (F/E)
17	Institutional and technical options in the development and management of small-scalle irrigation, 1998 (E)
18	Irrigation in Asia in figures, 1999 (E)
19	Modern water control and management practices in irrigation – Impact on performance, 1999 (E)
20	El riego en América Latina y el Caribe en cifras/Irrigation in Latin America and the Caribbean in figures, 2000 (S/E)
21	Water quality management and control of water pollution, 2000 (E)

Availability: November 2000

Ar	–	Arabic	Multil –	Multilingual
C	–	Chinese	*	Out of print
E	–	English	**	In preparation
F	–	French		
P	–	Portuguese		
S	–	Spanish		

The FAO Technical Papers are available through the authorized FAO Sales Agents or directly from Sales and Marketing Group, FAO, Viale delle Terme di Caracalla, 00100 Rome, Italy.